Neutron Stars, Black Holes and Gravitational Waves

Neutron Stars, Black Holes and Gravitational Waves

James J Kolata
University of Notre Dame, Notre Dame, IN, United States of America

Morgan & Claypool Publishers

ISBN   978-1-64327-422-5 (ebook)
ISBN   978-1-64327-423-2 (hardcover)
ISBN   978-1-64327-419-5 (print)
ISBN   978-1-64327-420-1 (mobi)

DOI   10.1088/2053-2571/aafb08

Version: 20190401

IOP Concise Physics
ISSN 2053-2571 (online)
ISSN 2054-7307 (print)

A Morgan & Claypool publication as part of IOP Concise Physics
Published by Morgan & Claypool Publishers, 1210 Fifth Avenue, Suite 250, San Rafael, CA, 94901, USA

IOP Publishing, Temple Circus, Temple Way, Bristol BS1 6HG, UK

*To my wife Ann for her constant encouragement
during the course of this book project.*

Contents

Preface

A little over a century ago, Albert Einstein published his General Theory of Relativity (GTR). Originally conceived as an extension of his Special Theory of Relativity (STR) to systems that are accelerating relative to each other, it turned out to be a new theory of gravity. Since gravity, perhaps counterintuitively, is actually a very weak force, measureable differences between the predictions of Newton's theory of gravity and the GTR only appear in extreme astrophysical scenarios. Over the years, the GTR has been verified again and again, most interestingly in the discovery of 'black holes', objects that are so massive that even light can't escape from their surfaces. However, one never-verified prediction was that of gravitational radiation. This was the idea that gravity, under certain circumstances, could propagate through space as a wave, similar in some respects to the electromagnetic radiation we know as light. The impetus for this book was the very recent (2015) discovery of gravity waves.

After brief discussions of the most relevant properties of waves, the STR, and the GTR, the evolution of massive stars to produce exotic objects such as 'neutron stars' and black holes is described. The properties of black holes are more extensively discussed since these provide testing grounds for the GTR. Next, the history of attempts to directly observe the elusive gravity waves, and the technological advances that ultimately led to their detection, are related. Finally, the fact that this discovery has led to the formation of the new science of gravity wave astronomy is detailed. The text contains many links to websites that extend and clarify the discussion. By following these, the reader can obtain a much more in-depth understanding of many of the concepts introduced here.

Acknowledgements

I acknowledge with gratitude the assistance of Jeanine Burke, Karen Donnison, and Melanie Carlson in the production of this manuscript in its final form.

Author biography

James J Kolata

James J Kolata is an emeritus professor of physics at the University of Notre Dame in the United States of America and a Fellow of the American Physical Society. He is the author of over 300 research publications in nuclear physics, as well as the book *Elementary Cosmology: From Aristotle's Universe to the Big Bang and Beyond,* which is also a part of the IOP Concise Physics Series.

Chapter 1

Introduction

In February of 2016, the world of physics and astronomy was shaken by a truly remarkable report in the journal *Physical Review Letters*: 'Gravitational Waves' had been directly detected. Predicted by Albert Einstein's General Theory of Relativity (GTR), published one hundred years earlier in 1916, gravitational waves were only the latest of many fantastic phenomena stemming from this ground-breaking work. The actual observation, by groups in the United States of America and Italy, actually occurred in September of the previous year. The intervening 5 months were used to make extensive checks of the observation in order to ensure that it wasn't, for example, a result of some failure in the detection equipment or any other anomaly. This is important for any scientific observation, but especially so in this case because observation of gravitational waves had been reported nearly 50 years earlier in the same journal. That report was ultimately shown to be wrong, both experimentally since it couldn't be reproduced by subsequent observations and theoretically since the magnitude of the observed effect was far too great to be explained in the context of Einstein's theory. Nevertheless, it generated a 50 year quest to actually observe these predicted waves using more and more sensitive instruments, which finally succeeded. Since then, five or six more such gravity-wave events have been seen and confirmed. One of them was located with a high enough degree of accuracy that its source could also be observed with optical telescopes, resulting in an incredible amount of information on topics as diverse as gamma-ray bursts and the synthesis of the chemical elements in the Universe.

The fact that it took so long for the quest to be fulfilled testifies to the extreme weakness of the expected signals, due both to the astronomical distances to the objects creating them and, as we shall see, to the relative weakness of the force of gravity itself. In fact, it appears that events strong enough to be detected at the present time can only be generated by the interactions of very exotic objects—'neutron stars' or 'black holes'. Both are extremely massive objects that are the end stages of the life of certain stars, although black holes may also result from the merger of many, perhaps billions, of stars in dense regions, such as at the center of the Milky Way (and likely most other galaxies). Black holes themselves are one of the predictions of the GTR, though the concept was actually first proposed nearly 150 years before Einstein developed his theory. In the following chapters, we'll describe both of these exotic objects, demonstrate how their interactions can lead to gravitational waves that can be detected on Earth, and clarify the nature of the expected signals. In order to do this, it will first be necessary to discuss the GTR, its predictions of neutron stars and black holes, the history of attempts to observe gravity waves, and the development of modern instrumentation that was finally sensitive enough to observe gravitational radiation. Lastly, we'll show how this new capability has expanded the range and scope of traditional optical astronomy and in the process ushered in a new age of 'multimodal' astronomy. This exciting new discipline opens up the potential for seeing objects and phenomena in the Universe that had heretofore been nearly invisible.

Chapter 2

Waves

Most of us are familiar with the concept of waves. Imagine throwing a stone into a placid pond of water. A circular disturbance consisting of 'crests' and 'troughs', places where the water is higher or lower than the surface of the pond, moves out from the location where the stone hit (figure 2.1).

Two of the more interesting phenomena associated with waves are 'diffraction' and 'interference'. As an example of the first of these, consider what happens when a water wave goes through a small opening in a breakwater (figure 2.2). In the image on the left, the waves hitting the breakwater were generated at a very large distance from it, and therefore the wave crests look like straight lines. These are known as 'plane' waves. As the wave moves through the small opening, however, it spreads out into a circular pattern. This is diffraction. Now suppose that there are two closely-spaced openings in the breakwater, as in the image on the right. Where the waves diffracted through the two openings interact, their crests will sometimes line up, producing a bigger disturbance. This is called 'constructive' interference. On the other hand, in other locations, the crest of one aligns with the trough of the other and the result is a minimum (or no) disturbance. This is known as 'destructive' interference. The pattern on the wall where the waves in this image hit is a sequence of high and low disturbances.

Diffraction and interference are characteristic of waves, and Thomas Young used this to show, in 1804, that light was a wave phenomenon (figure 2.2). When a beam of light goes through a single slit, a fuzzy image of the slit is produced on the wall behind it; fuzzy because the image is spread out by diffraction (figure 2.3). However, when the light is allowed to go through two closely-space slits, the image is replaced by a series of bright spots recording the positions where constructive interference occurs.

Once it was determined that light was a wave phenomenon, it was natural to ask just what was waving. In the case of waves on a pond, the 'medium' of the waves is water. What is the medium of light waves? Since light travels between the Sun and

Figure 2.1. Waves on water. Reproduced with permission from CC BY-SA 4.0.

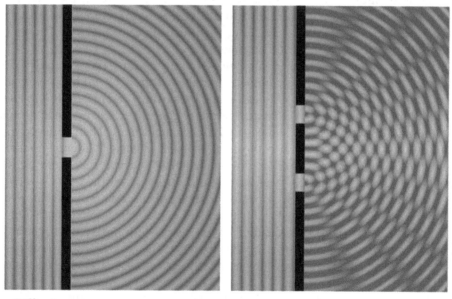

Diffraction   *Vector   EPS 10*

Figure 2.2. Diffraction (left) and interference (right) of waves. By Fouad A Saad. Shutterstock.com.

the Earth, it was at first assumed that there must be some unknown medium (called the 'aether') that pervades even the vacuum of space. The American physicists Michelson and Morley (M&M) reasoned that light from a distant star would encounter an 'aether wind' due to the motion of the Earth, and they designed an apparatus to measure it. Their idea was that the aether wind was similar to the current in a river. It turns out that a boat traveling directly across a river and back, actually traveling at an angle to compensate for the current, always arrives earlier

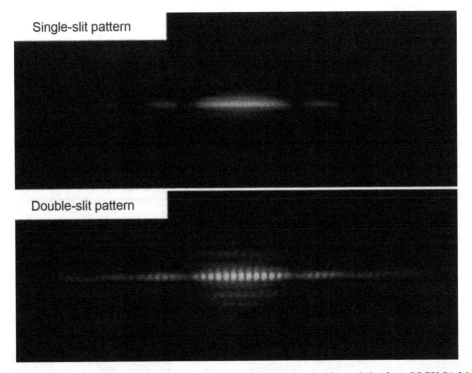

Single-slit pattern

Double-slit pattern

Figure 2.3. Single- and double-slit patterns observed with light. Reproduced with permission from CC BY-SA 3.0.

than a similar boat traveling directly upstream and then back to the starting point, (assuming that both boats move at the same speed relative to the water). In M&M's experiment, a light wave was divided in two using a partially-silvered mirror (figure 2.4). One of these waves traveled perpendicular to its initial direction and was then reflected backward to a detector. The other wave continued in the original direction, was reflected back by another mirror, and then was reflected at 90° by the partially-silvered mirror so that it entered the same detector:

If the original direction was perpendicular to the 'aether wind', then the second wave traveled across the stream and should arrive at the detector first. The time difference could be measured with high accuracy by the interference phenomenon. For example, if the later-arriving wave came one full wavelength (the distance between consecutive crests) late, the peaks of the two waves would coincide and add together to produce a bright spot. On the other hand, if it came one-half wavelength late, its peak would fill the valley of the earlier wave and the result would be a dark spot due to 'destructive interference'. Since the speed of light is very large and its wavelength very small, M&M's apparatus was capable of measuring exceedingly small time differences. This was important since they knew that the speed of light is 10 000 times greater than that of the Earth in its orbit so the expected effect was tiny. Of course, it wasn't really possible to ensure that the apparatus was oriented perpendicular to the aether wind, so it was designed to rotate around its axis. Measurements were made at many different angles. In addition, data were taken at

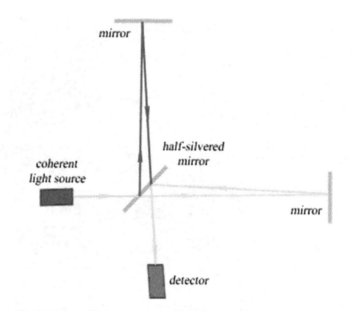

Figure 2.4. The Michelson–Morley experiment. Reproduced with permission from CC BY-SA 3.0.

intervals of 1/2 year so that the direction of the Earth relative to the aether presumably changed. This would increase the expected effect. Yet, despite all their care, they were never able to observe any effect at all, as they reported in 1887. Subsequent attempts to repeat the experiment with improved sensitivity gave identical results: there is no aether wind and therefore no aether and the speed of light is always the same regardless of the motion of the Earth. This conclusion provided very important information that Albert Einstein eventually used to construct his Special Theory of Relativity.

Chapter 3

The Special Theory of Relativity

These days, most people are familiar with Albert Einstein's famous equation relating energy (E) and mass (m):

$$E = mc^2$$

Here, the quantity c^2 is the speed of light squared. This equation is interpreted as follows. In addition to any kinetic energy of motion an object might possess, there is an additional term, equal to the mass of the object times the speed of light squared, which appears even for stationary objects. This is called the **rest-mass energy**, and it's very, very big because the square of the speed of light is a huge number. Apart from this term, the kinetic energy of motion at a low speed v approaches the value of $mv^2/2$ given by Isaac Newton (as it must) but behaves quite differently at higher speeds. The equation for the rest-mass energy implies that mass can be converted into energy (and energy into mass), with world-shattering consequences, such as the 'atomic' bomb. It also transformed our understanding of the entire universe by (as only one example) providing an explanation for the vast amounts of energy emitted by our Sun.

This iconic equation is the most famous result from the Special Theory of Relativity. The theory is 'special' because it's restricted to the case of an object (or person) moving at a **constant speed** relative to another object. (In other words, the effects of accelerated motion are not included. That is the subject of the General Theory of Relativity to be discussed later.) Its genesis was Einstein's fascination with the idea of light, and in particular with what someone might observe if he or she were moving at a speed near **c**. Light was known to be a wave phenomenon, so an analogy might be to a surfer riding the crest of an ocean wave. To a person on the shore, the wave would be moving, but to the surfer it would appear to be stationary. However, when applied to light, the idea of a stationary wave doesn't seem to make sense since our eyes respond to the motion of the light wave. Does light just disappear for people moving at high speed? Furthermore, the Michelson and Morley

3-1

experiment discussed in the last section had shown that the speed of light is always the same regardless of the motion of the Earth. Finally, and most importantly, there was the fact that the Scottish physicist James Clerk Maxwell published in 1862 a set of equations that today forms the basis of the science of electricity and magnetism. He showed that these two apparently different phenomena were actually just two manifestations of the same effect: electromagnetism. Remarkably, his equations also predicted the existence of an 'electromagnetic wave' which requires no 'waving' medium and moves at a speed that could be calculated from two well-measured constants relating to electricity and magnetism. This calculation gave a speed of 3×10^8 m s^{-1}, exactly equal to the measured speed of light. (This is actually the speed of light in a vacuum, as it's known to travel slower through other media.) There is no reference in Maxwell's theory to either a source or an observer. The speed of light through a vacuum is therefore a universal constant, independent of the motion of either the source or the observer. That's exactly what's required to explain Michelson and Morley's result, but this concept was very difficult for physicists of the late 19th century (and for most people) to accept. To understand why, consider the speed of a baseball thrown from a moving truck as measured by someone standing at the side of the road. Suppose the truck is moving at 60 miles per hour and the baseball is thrown in the direction of motion at 50 miles per hour relative to the truck. Then, a stationary observer at the side of the road will presumably measure a speed of 110 miles per hour relative to the road; the sum of the speeds of the truck and the baseball. On the other hand, if the baseball is thrown backward, its speed relative to the road is only 10 miles per hour, the difference of the two speeds. (These calculations neglect the effect of air friction.) This result seems to be intuitive based on our experience of how things move at low speed relative to **c**. Light must apparently be very different from baseballs in some mysterious way since its speed is always the same: light from a moving flashlight has the same speed as if the flashlight were stationary.

Einstein's great contribution was to postulate the constancy of the speed of light as required by Michelson and Morley's experiment and Maxwell's equations, and then derive a new theory of motion based on this principle. Now the speed of an object is d/t, the distance it moves in a time t divided by t, so he reasoned that there might be a problem with our understanding of the nature of time. Since the age of the Greek philosophers (~400 BC), and even before, it had been assumed that time exists independently of any observer and progresses at a consistent rate throughout the Universe. This immutable nature of time might seem obvious based on our everyday experience, but is it really always true? After all, the behavior of time at speeds near that of light had never been directly measured. Since time is measured by clocks, we need a 'light clock' to probe what happens at very high speeds (figure 3.1). One way to construct such a clock, at least in principle, is with a flash bulb that sends a pulse of light toward a mirror. The light pulse then bounces back to a detector and we count one 'tick' of the clock as the time it takes for the flash to be detected. It is necessary to use light in this 'thought experiment' because its speed is not affected by the motion of the clock according to Einstein's postulate. Suppose we now place such a clock on a truck and orient it so that the light pulse travels perpendicular to

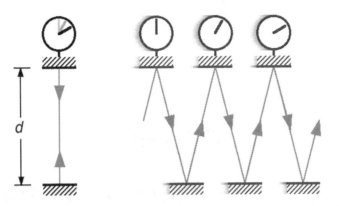

Figure 3.1. Light clock as seen by an observer moving with it (left) and one watching it move (right). Reproduced with permission from CC BY-SA 3.0.

the bed of the truck as in figure 3.1. A person on the truck sees the light travel straight up and down, no matter if the truck is moving or not, and so the ticking of the clock is unaffected by the motion of the truck according to this observer.

The crucial question is: **what is seen by a person standing on the side of the road?** According to this second observer, the light beam doesn't travel straight up and down since the truck moves while the clock is ticking. Instead, it travels toward the upper mirror on the hypotenuse of a right triangle and then back down again along a similar path, travelling a longer distance as shown in the right-hand portion of the figure. Since the speed of the light beam seen by both observers must be the same, the person at the side of the road has to conclude that the clock takes a longer time to tick. In other words **'moving clocks run slow'** and time is not an absolute quantity but instead is relative to the motion of the observer. This effect has come to be called 'time dilation' since it implies that intervals measured by a moving clock are stretched out. Einstein's insight was that time dilation is not specific to light clocks but rather applies to all clocks (including biological clocks). The effect has now been verified many times with real clocks moving at high speed. No matter how bizarre this may seem, it represents the true nature of time itself. We are unaware of this very fundamental fact of the Universe only because the effect is vanishingly small unless one travels at speeds near to that of light.

With this insight, Einstein was able to derive an entirely new theory of motion using only simple mathematics such as basic algebra or, in the case of the energy equation, rather elementary calculus. In this theory, time is not absolute but instead behaves like a fourth dimension, on an equal footing with the familiar three spatial dimensions, and we live in a **four-dimensional spacetime**. These two concepts, spacetime and the interconvertibility of mass and energy, play an important role in any attempts to understand the Universe around us.

Neutron Stars, Black Holes and Gravitational Waves

James J Kolata

Chapter 4

The General Theory of Relativity

Einstein's General Theory of Relativity, published in 1916, extended the study of relative motion to the case of accelerating objects. Unlike the case of the Special Theory, which requires the acceptance of the counterintuitive notion that time is a fourth dimension and not an absolute quantity, it's possible to achieve some understanding of the nature of accelerated motion based on experiences that are probably familiar to everyone. When you ride in an elevator that begins to accelerate upward, you feel heavier due to this acceleration. The increase in weight is real and can actually be measured if you're standing on a scale. When the elevator begins to accelerate downward, the result is a measurable decrease in your weight. If the cable is cut (and there are no brakes), you experience 'free fall' and your weight measured on the scale goes to zero. This is the exact situation for astronauts in the International Space Station, who are always freely falling as the station orbits the Earth. If the elevator is either stationary or moving at a constant speed, your weight returns to its normal value resulting from gravitational attraction towards the center of the Earth (figure 4.1). So far, so good.

But now suppose that the elevator is moved to some location in space very far away from any planet or star. The force of gravity, and therefore your weight as measured on the scale, becomes negligibly small. If it now accelerates upward (in other words, toward the top of the elevator) at exactly the acceleration due to gravity on the surface of the Earth, your weight measured on the scale will be its normal value. If you can't see outside, there's no way for you to distinguish the latter case from that of a stationary elevator on the surface of the Earth. In other words, 'gravity and acceleration are equivalent'. This is a statement of Einstein's equivalence principle and forms the basis of the General Theory, which is actually a theory of gravity. The problem with it is that, unlike the case of the Special Theory, the mathematics required turns out to be exceptionally difficult. In fact, even now the relevant equations can only be solved in a relatively small number of situations.

doi:10.1088/2053-2571/aafb08ch4 4-1

Figure 4.1. An elevator on Earth (left) and in a rocket ship (right). Reproduced with permission from CC BY-SA 3.0.

The accepted theory of gravity in 1916 (and even today in most cases) is due to Isaac Newton. Einstein compared Newton's second law (F = ma) and gravity equation (F = GmM/R^2). Here, **F** is the force of attraction between two masses, **a** is the acceleration due to gravity, **R** is the distance between the two masses, and **G** is a universal constant. The effect of gravitational attraction on an object of mass m attracted to an object of mass M is calculated by equating these two forces. In the process, the quantity m cancels out since it appears on both sides of the equation. This shows that **gravitational attraction is independent of the mass of the accelerating object**. Why is this? The m in F = ma is called the 'inertial mass' since it measures the resistance to change of motion. The m in the gravity equation is the 'gravitational mass' which measures the response to gravitational forces. Apparently these are equal. In fact, experiments have been carried out that show that they are exactly equal to better than one part in a trillion! Surprisingly, there was no compelling theoretical reason why this should be the case. The second thing to note is that Newton never really described how gravity, with its spooky action at a distance, actually worked. Finally, one important result from the Special Theory of Relativity is that nothing, including gravitational attraction, can move through spacetime faster than the speed of light. As just one example, if the Sun were to suddenly disappear, the Earth's orbit would be unaffected until 8 min later. Newton's theory, in contrast, assumes instantaneous action at a distance. With this background in mind, Einstein's insight was that the effect of the central mass in the gravity equation is to distort spacetime around the object, forming a 'gravity well' (figure 4.2):

The orbiting mass is simply following the path of least resistance (called the '**geodesic**') in the local spacetime. If there were no central object, spacetime would be

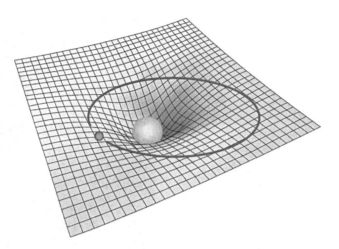

Figure 4.2. Orbit of a planet in a 'gravity well'. Adapted from https://en.wikipedia.org/wiki/File:Spacetime_curvature.png.

flat and the mass would proceed in a straight line. Warping of spacetime results in an orbit. A convenient demonstration of this is provided by a rubber membrane stretched over the surface of a barrel. A ball rolling along this surface will proceed in a straight line. However, a heavy mass placed at the center of the membrane will deform it and the ball will roll in an orbit as in the illustration above. This concept solves the problem of action at a distance: the mass is simply following the geodesic in local spacetime. It also explains the fact that the path depends only on the mass of the central object, which determines the amount of curvature. The theory can be shown to be identical to Newton's theory of gravity in the limit of weak gravitational forces. This is usually the case, so that Newton's equation is sufficient for most applications.

There were three 'classical' tests of general relativity, which it passed successfully. The first was a measurement, during a solar eclipse in 1919, of the bending of light from distant stars as the light passed close to the Sun (figure 4.3); this causes a distortion in the apparent location of the stars that was identical to the predicted value. (Actually, Newton's theory also predicts such a shift but of a smaller amount.) The second test had to do with shifts in the orbit of the planet Mercury. Such shifts can be caused by attraction to the other planets, but the value calculated with Newton's theory was too small in the amount of 43 out of 574 s of arc per century. (One second of arc is 1/3600 of a degree.) This was a small but worrisome discrepancy, but general relativity exactly predicts it. The third test has to do with the fact that spacetime is being warped so gravity also affects the flow of time. Calculations showing that time slows down in an intense gravitational field were confirmed by the 'Pound-Rebka' experiment in 1959. This results in '**gravitational redshifts**' (or blueshifts) distinct from those predicted by special relativity. Since then, the theory has successfully passed many more tests.

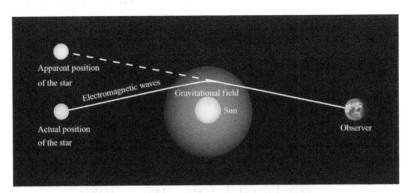

Figure 4.3. Bending of starlight in a gravity well.

4.1 Gravitational radiation

Einstein showed that his general relativity equations predicted the possibility of the emission of gravity waves, just as Maxwell's equations predicted light waves. Both travel at the speed of light. As one possible illustration of this phenomenon, consider the image above showing the warping of the **space–time continuum** by a star, generating a gravity well. If the star changes its shape in a regular fashion (as some stars are known to do), spacetime near its surface is disturbed. Ripples are formed and radiate outward in analogy to the emission of sound waves from a vibrating speaker. Because the gravitational force is actually very, very weak, this type of event does not produce gravity waves of sufficient intensity to be detected on Earth, even from a star as close as our Sun. To begin to appreciate this, consider that electromagnetism is approximately a trillion-trillion-trillion times stronger than gravity! There are two reasons why it nevertheless seems to us that gravity is stronger. First of all, most matter in the Universe is 'electrically neutral'. It contains equal numbers of positive and negative charges so that the electromagnetic effects cancel out and we only become aware of their strength when, for example, a (very small) fraction of atoms become 'ionized', with unequal numbers of positive and negative charges, as in a lightning storm. Secondly, the force of gravity is always attractive so the attraction from each of the approximately trillion-trillion atoms in a small sample of matter is additive. More catastrophic astrophysical events, such as the collapse of a supernova forming a 'neutron star' or 'black hole' or even the merger of two such objects, are therefore required to generate a detectable gravitational wave. Even then, the size of the signal is very small, which accounts for the fact that gravity waves have only recently been detected.

Chapter 5

The life of massive stars

Stars are created via the collapse of interstellar dust and gas clouds under their mutual gravitational attraction. The clouds heat up as they are compressed until 'fusion' of hydrogen begins to occur. This process results in the formation of the element helium through a chain of nuclear reactions, but the mass of the helium product is less than the mass of its constituent particles. The 'missing mass' is converted into energy according to the $E = mc^2$ relationship. The energy streaming from the center of the cloud eventually stops the collapse of the proto-star, which then is said to have reached 'hydrostatic equilibrium' (figure 5.1).

The subsequent history of the star depends critically on its mass. Very massive stars need to be very hot in order to reach equilibrium, and nuclear fusion occurs much faster in them because of this. Eventually, the hydrogen gas in their cores is entirely fused into helium, within only about 50 million years for a star that has 25 times the mass of our Sun. While the same thing will happen to our Sun, its expected lifetime is about 10 billion years. At this point, hydrogen-fusion energy disappears and the star again begins to collapse under gravitational attraction until the temperature at its core rises sufficiently for helium to fuse. (Helium fuses at a higher temperature than hydrogen because it has two electrical charges rather than one, so the electrostatic repulsion that must be overcome for the helium nuclei to get close enough to fuse is greater.) When helium fusion begins, the star again reaches hydrostatic equilibrium but its outer atmosphere expands and cools, forming a 'red giant' star. In the case of the Sun, its surface will then be somewhere between the orbits of the Earth and Mars. The red-giant phase lasts about 100 million years for the Sun (1 million years for a 25 solar mass star) until the helium fuel is used up. At this stage, the history of low-mass versus high-mass stars diverges. The gravitational mass of the Sun is too small to ignite fusion of the heavier elements, such as carbon, nitrogen, and oxygen formed from helium fusion. The Sun will nevertheless continue to contract and heat up until the atoms in its core are packed so tightly together that repulsion of their electron clouds stops the collapse. In this process, its outer layers

Figure 5.1. Hydrostatic equilibrium in a star.

are completely ejected, forming a 'planetary nebula' with an extremely hot 'white dwarf' star at its center. In normal matter under conditions found on the surface of the Earth, there is a lot of empty space between constituent atoms and the density of all materials is relatively small, less than 25 times that of water at most. However, the atoms in the core of a white dwarf are basically touching each other and the relevant density is about 350 000 times that of water! Over many billions of years, the white dwarf will cool and eventually disappear from view.

The 25 solar mass star will also go through the helium-fusing stage, forming a red 'super-giant' star, but will run out of helium fuel in its core in only about 1 million years as mentioned above. Big stars live fast and die young! It also has a very different fate in store for it, since such stars have enough gravitational mass to ignite fusion of successively heavier elements. In a relatively short time, it will fuse first carbon, then magnesium, then silicon, and finally sulfur to form iron. (This sequence has been simplified but is basically correct.) According to calculations, the resulting object will have an iron core with layers of successively lighter elements arranged in

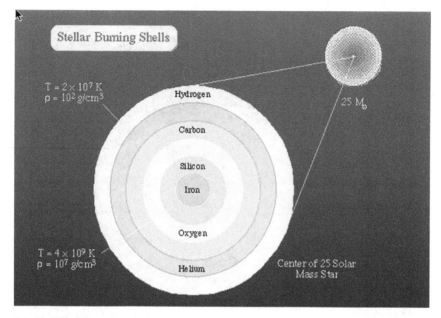

Figure 5.2. Shells in a 25 solar mass supernova precursor. Adapted from https://commons.wikimedia.org/wiki/File:Evolved_star_fusion_shells.svg.

an onion-like structure as in this figure 5.2. (Note that the term 'burning' in this figure actually refers to nuclear fusion.)

The star now has a serious problem. According to well-understood laws of nuclear physics, fusion of iron does not generate energy. Instead, it acts like a refrigerator instead of a furnace. The star, which once was stabilized by the energy of fusion, now has no way to prevent a rapid collapse under the influence of gravity. It first implodes in a very short time until it reaches white-dwarf density, then bounces back and is destroyed in a supernova explosion. The elements 'cooked up' in its various fusion stages are ejected into the surrounding interstellar medium, perhaps to be incorporated into later generations of stars. This is the case for our Sun, which has been shown to be at least a 'second generation' star because it contains significant amounts of heavier elements, such as carbon, nitrogen, and oxygen, that could only have been produced in the 'helium-burning' stage of stellar evolution. (The Sun will not enter that stage for another 5 billion years or so.) Observations of supernova remnants have detected all the elements, including iron, that result from the various fusion stages. Calculations have also shown that some elements heavier than iron are formed in the very energetic supernova explosion itself. The exact details of this process are the subject of ongoing research in nuclear astrophysics. Moreover, at least a portion of the inner core of a supernova might escape dispersion, resulting in the formation of more exotic objects as discussed further below.

The fates of stars having masses between 1 and 25 times the mass of the Sun have also been the object of extensive astrophysical research. It has been shown that, for stars with masses less than 1.44 solar masses (called the Chandrasekhar limit), there is not enough gravitational energy to proceed beyond the white dwarf stage. However, it turns out that stars with initial masses (prior to collapse) of roughly 8 solar masses will end up as white dwarves anyway since considerable mass is ejected during the collapse itself.

5.1 Neutron stars

Compression of the matter in the supernova core can on occasion push through the repulsion of the electron clouds, forming a state of matter in which the atomic nuclei are packed tightly together and the electrons are pushed back into the protons to form a 'neutron star'. This object has a density of about 500 trillion times that of water, corresponding to the entire mass of the Sun compressed into only a 10 km radius! Because of the physical law of conservation of angular momentum, the star's rate of spin increases dramatically during the collapse, up to as much as 40 000 revolutions per minute or more. Also, if like the Sun the progenitor star had a significant magnetic field, the compression would cause it to increase, typically to something like a magnetic field at its surface that is at least a trillion times that of the Earth. Neutron stars having magnetic fields even hundreds of times larger than this, called magnetars, have been detected. These fields are so powerful that they could kill a person from 1000 km away by warping the atoms in living flesh! However, they pose a threat at much greater distances than this because of the powerful flares they eject. Due to their very large magnetic fields, neutron stars are relatively easy to detect because they form what are known as 'pulsars'. The first of these was discovered in July of 1967 by Jocelyn Bell Burnell using a radio telescope. She found that the signal from a peculiar object pulsed on and off in a very regular pattern, at a rate of approximately one pulse per second. It was initially called LGM-1 (for 'little green men') because of the suggestion that the signal might be an attempt at communication from intelligent inhabitants of a distant star. However, it was soon realized that this was not the case. The explanation for the regular pulsation was more prosaic but almost as interesting. Flares emitted from the surface of a neutron star are channeled through its magnetic poles, just as the charged particles emitted by our Sun during a solar flare are directed by the Earth's magnetic field to form the Aurora Borealis and the Aurora Australis when they strike the Earth's atmosphere. In the case of neutron stars, these flares emit radio waves that are channeled along the magnetic poles. If, as is also the case for the Earth, the magnetic poles are not exactly aligned with the geometric poles of rotation, then the corresponding radio emissions rotate in space at the frequency of rotation of the neutron star. If the Earth happens to be in a location such that these radio waves sweep past it, then the result is a pulsing emission much like the flashing light you see from a lighthouse (figure 5.3).

The discovery of pulsars was awarded the 1974 Nobel Prize in Physics to Anthony Hewish and Martin Ryle, but Jocelyn Bell, who was at the time a graduate student in the laboratory of Anthony Hewish, controversially did not share in the prize. However, in September of 2018 Jocelyn Bell Burnell, professor of astronomy at

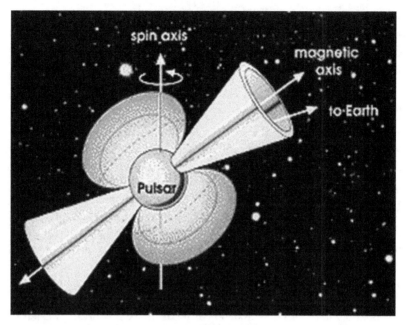

Figure 5.3. The 'lighthouse effect' applied to pulsar radiation. Image: NASA.

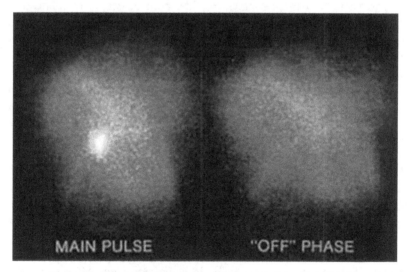

Figure 5.4. Pulsed x-ray emission from the center of the Crab Nebula.

Oxford University, was awarded the Breakthrough Prize in Fundamental Physics for her discovery. Previous awardees included such luminaries as Stephen Hawking. A considerable number of pulsars have now been found. One example is the pulsar in the Crab Nebula, the remnant of a supernova observed by Chinese astronomers in the year 1054. An x-ray image of its center during and between pulses (which occur 33 times per second) is shown in figure 5.4.

Pulsars are very accurate clocks. In fact, the most accurate clock in the world (at the time) was the 'pulsar clock' installed in Gdansk, Poland, in 2011. However, they have been observed to slow down over very long periods of time due to the emission of electromagnetic energy resulting from their rotating magnetic fields. For example, the period of the Crab pulsar is slowing by approximately 12 microseconds per year. In certain cases, this slowdown may also occur due to the emission of gravity waves, as will be discussed further below.

Chapter 6

Black holes

If the mass of a neutron star exceeds about 2.5–3 solar masses, then its gravitational pressure will be so great that even neutrons can't withstand it. (The exact value is not very well known, but the heaviest neutron star so far observed has a mass of 2.3 solar masses.) Under these conditions the surface of the neutrons is breached and the collapse continues catastrophically, forming a 'black hole', which is a region of spacetime from which gravity prevents anything, even light, from escaping. The possibility that this might happen was discussed by John Michell in 1783 and at about the same time by Pierre Laplace, but the first solution of Einstein's General Theory of Relativity showing that these objects can exist was published by Karl Schwarzschild in 1916. The principle can best be understood in analogy with what happens to objects thrown upward from the surface of the Earth. As we all know, a baseball thrown directly upward will slow down under the influence of gravity, eventually stop, and then fall back to Earth. However, a rocket ship traveling fast enough (at the 'escape velocity' of about 25 000 miles per hour) will break free from the Earth's gravity and leave without ever falling back to Earth. It turns out that the square of the escape velocity from the surface of an object is dependent only on the ratio of the mass of the object to its radius. If the mass M is large enough or the radius R small enough (or both), then the escape velocity will equal or exceed the speed of light and not even light can escape from the surface of the object. Since the Special Theory of Relativity states that nothing can travel through space faster than light, nothing can escape the gravitational pull at the 'surface' of a black hole. Schwarzschild calculated the so-called 'Schwarzschild radius' R_s of a (non-rotating) black hole of mass M. R_s for the Earth turns out to be only 9 mm (about 1/3 of an inch). If some processes were to compress the entire Earth into a ball of this radius, it would become a black hole. The more massive the black hole, the bigger it is. However, the Schwarzschild radius is not like the radius of a ball bearing. All of the matter within R_s falls into the center of the black hole, forming a so-called 'singularity'. The Schwarzschild radius only represents the boundary from within which nothing can escape, called the 'event horizon'.

6.1 Some properties of black holes

Since no light, and therefore no information, can come from inside R_s, only quantities that are 'conserved' according to the laws of physics can remain after the black hole is formed. In particular, these are the mass, charge (if any) and angular momentum (rotation) of the matter within its surface. This is known as the 'no-hair theorem' and it implies that a black hole cannot have a magnetic field. If a massive star prior to its collapse possessed a magnet field, that field will be radiated away as electromagnetic radiation in a short period of time during the collapse. This makes it quite difficult to detect a black hole resulting from a supernova explosion, unlike the case of a neutron star. Only a few have been detected, and they are in binary systems where matter from a neighboring star is pulled into a black hole generating a lot of energy. An example is the object known as Cygnus X-1, which was the first black hole detected (in 1972) as one of the very strongest x-ray sources ever detected on Earth (figure 6.1). It's a 'blue-giant' normal star orbiting a compact object with a mass of approximately 15 solar masses (determined by analyzing the orbit). The radius of the companion has been shown to be too small to be anything but a black hole. (However, it hasn't been directly imaged.) The artist's impression of the system below shows infalling matter from the large star that, according to calculations, forms an 'accretion disc' before spiraling into the black hole, emitting x-rays in the process.

However, there is another way in which black holes can form. The density of stars in the center of a galaxy is very high and they can collide and merge, eventually forming a 'supermassive' black hole. In the case of the Milky Way, observations have shown that there exists a black hole at its center with a mass about 4 million times that of the Sun. Analysis of the motion of stars around this object, known as Sagittarius A*, have confirmed its mass and radius demonstrating that it must be a black hole. Supermassive black holes having masses up to billions of times that of the Sun have been detected in other galaxies from the energetic jets that emerge from them. An example is the Galaxy known as Cygnus A shown in figure 6.2. It is thought that every galaxy may well have a supermassive black hole at its center.

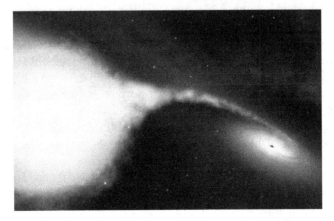

Figure 6.1. An artist's impression of the Cygnus X-1 binary system. ESA/Hubble illustration.

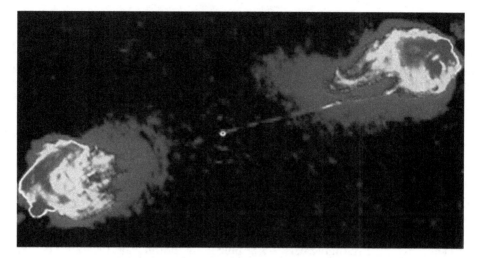

Figure 6.2. Cygnus A. The galaxy is the small dot at the center, which emits huge plumes. Image courtesy of NRAO/AUI.

The (Schwarzschild) radius of a black hole increases when matter falls into it to account for its greater mass. It also turns out that the 'gravitational tidal effects' at the event horizon become weaker as the mass and radius of the black hole increases. To understand these tidal effects, think about the effect of gravity on your body as you stand on the surface of the Earth. Since gravity is slightly weaker at your head than at your feet, there is a very small net force tending to pull you apart. This effect is greatly magnified near a solar-mass black hole due to its much greater gravitational attraction, but it is smaller for a supermassive black hole due to its enormous size. This means that you could in principle cross the event horizon of a super-massive black hole, such as the one at the center of our galaxy, without immediately being torn apart by tidal forces, and you might in fact not even notice that you had actually passed the point of no return!

An interesting sidelight of falling into a supermassive black hole concerns the nature of this event as observed by you (as you fall into the black hole) and someone at a distance from the black hole. You would observe nothing special on crossing the event horizon, and it would take only a short while (as measured by your clocks) for you to fall into the 'singularity' at its center. There is no way to avoid this. Once you cross the event horizon, you are doomed to be compressed into the singularity. However, because of 'gravitational time dilation', an observer at a distance from the black hole would see you approach the event horizon but never cross it! This is one of the strange effects of the General Theory of Relativity.

The rotation rate of a black hole has a natural limit, when the event horizon is rotating at the speed of light. This means that the larger (i.e. more massive) the black hole, the slower its maximum rate of spin. It turns out that the event horizon of a non-rotating black hole is spherical, which prevents us from seeing the singularity at its center. This has become known as 'cosmic censorship'. However, the event horizon of a rapidly spinning black hole is deformed and bulges at the equator (just like the Earth). This brings up the interesting question if the deformation could ever be great enough that we could directly observe the 'naked singularity' at its center. This question has not been answered yet. It was the subject of a famous bet between Steven Hawking and Kip Thorne. The rotation of a black hole is interesting for another reason. A solution of Einstein's general relativity equations for this case (the 'Kerr solution') reveals that the rotation of a black hole causes a 'swirling' of spacetime outside of it, i.e. outside the event horizon. Because of this, it turns out to be possible to extract a really huge amount of rotational energy from a spinning black hole (thus slowing down its rate of spin). This is thought to be the source of the energy that powers the immense jets of matter and energy emerging from galactic cores containing supermassive black holes. It is also presumed to be the source of the energy that powers 'quasars' distant 'quasi-stellar' objects that look like points of light but are very far away. Recently, galaxies surrounding some of these quasars have been observed. The energy output of these objects is variable on the time scale of days or shorter, so they must be very compact. Only emission from supermassive black holes can account for these properties. (Quasars are all very distant from us, so it is thought that they are the end result of the evolution of galaxies formed when the Universe was much younger... possibly the first galaxies to form.)

6.2 The thermodynamics of black holes

Stephen Hawking deduced some fundamental and important properties of the energetics of black holes from a consideration of a fundamental principle of physics, the second law of thermodynamics. In order to discuss this, we need to introduce the concept of 'entropy', which is a measure of the 'disorder' of a system. As an example, consider the number of different sequences that can result from flipping **n** coins. For **n** = 3, there are 8 possibilities, for **n** = 4 there are 16 possibilities, and for **n** = 100 there are 2^{100} possibilities. The entropy for a random toss of 100 coins is $\log(2^{100}) = 30.1$, where 'log' indicates the logarithm to the base 10. In this example, $10^{30.1} = 2^{100}$. Logarithms are used because one rapidly gets very big numbers for most systems, and large entropy means a large degree of disorder. The second law of thermodynamics states that **the entropy of an isolated system always increases**. That does not mean that disorder is inevitable. For example, I could choose to arrange the sequences of coin tosses in some predetermined order, thereby lowering its entropy. However, it takes energy (mental and physical) for me to do this, and the second law says that the total entropy of the system (me plus the coins) increases in this ordering process. My use of internal and external energy increases my entropy more than enough to compensate for the decrease in the entropy of the coins. Viewed in this way, the order of living beings on the Earth (as only one example) is made possible by energy from the Sun, which in the process increases the Sun's entropy. The second law is an extremely powerful tool, and it is closely bound up with the nature of time. For example, consider watching a movie of a glass bottle shattering under the impact of a rock. If you run the movie backwards, it looks strange. Why? Because the disordered glass fragments rearrange themselves into an ordered bottle, and this never happens by itself. The direction of time-flow is related to the increase in entropy. Another thing to consider: no physical principle other than the second law prevents all the air molecules in a room from gathering by chance in one corner, which would be a disaster for us, of course. In fact, this **can** happen by chance, but the probability it **will** happen is vanishingly small. The second law deals with the probabilities of such rare events. As such, it is different from physical laws such as Newton's laws of motion because these make absolute predictions. The second law is more akin to '**probabilistic**' theories such as quantum mechanics.

Jacob Bekenstein was one of the first to notice that there is a deep connection between the entropy of a black hole and the area of its event horizon. In fact, he showed that **the entropy of a black hole is equal to the ratio of this area to the square of the Planck distance**. For a 10 solar mass black hole, $R_s = 30$ km, $A = 2.8 \times 10^9$ m^2, and the entropy is 10^{79} (not 79… the log has already been taken)! This is an enormous amount of disorder for an object that is supposed to be simple. Remember: 'a black hole has no hair'. Initially, this was viewed as a problem, which was one of the things that retarded the study of black-hole thermodynamics. In retrospect, it is clear that this result is mandated by the second law. For example, we could take all the gas molecules in a room, which have very large entropy due to their random motions, and dump them into a black hole. In the process, a lot of entropy disappears from the neighborhood because all we can know about the air molecules within the black hole is their total mass (and electric charge and angular momentum if they had any). Their random motions are not visible anymore. The second law says that the total entropy of the Universe must increase, though, so the entropy of the black hole has to go up. In fact, it is now understood that **the entropy of a black hole is a measure of the number of different ways we can form it**, which is very large. This disorder must reside inside (or on the surface of) the black hole. According to Bekenstein's theorem the area of the event horizon, which increases in size as more mass is added, is a measure of this disorder.

6.3 Hawking radiation

We now come to Steven Hawking's insight. In 1974, he put it all together when he showed that not only was the area of its event horizon related to the entropy of a black hole, but also **the surface gravity at R_s was a measure of an 'effective temperature'** of the black hole. In the process, he showed that black holes can radiate not only photons but also particles of many kinds. This is now known as 'Hawking radiation'. How does it work? The process goes something like this: a quantum fluctuation produces, e.g. a 'virtual' electron–antielectron pair near the event horizon, which is then torn apart by the tidal gravity of the black hole. This makes the particles real by taking energy from the gravitational field. One of the pair then drops into the hole and the other radiates into space. Of course, the same thing can happen with virtual photons, so the hole also radiates electromagnetic radiation. In Hawking's words: '**black holes ain't black**'. What are the properties of this radiation? First of all, Hawking showed that **the temperature of a black hole is inversely proportional to its mass**. The temperature of a 4 solar mass black hole, for example, is 1.5×10^{-8} K. This is much less than the temperature of the cosmic microwave background, which means that in fact it is a net absorber of radiation from the CMB. All 'normal' black holes are heavier and therefore even colder than this. However, there may be an exception to the rule. Hawking speculated that 'primordial black holes' might have been made in the Big Bang and survived to the present. Hawking radiation implies a loss of energy (and therefore mass) from the black hole. But a smaller black hole has a higher temperature and the emission of radiation goes as the fourth power of the temperature, so the hole heats up as it shrinks and therefore loses mass even faster. How long does it take for the black hole to completely evaporate? It can be shown that the lifetime of a black hole (neglecting absorption of the CMB) is proportional to the cube of its mass. For a 10 solar mass black hole, it works out to 2×10^{78} s. The present age of the Universe is about 5×10^{17} s so we don't have to worry about evaporation of 'normal' black holes. However, one can ask for the mass of a black hole that lives for the current age of the Universe. The answer turns out to be 1.3×10^{11} kg, corresponding to a radius of about 20% of that of a proton. Now, the Hawking process is a prescription for a 'thermal runaway' in which most of the energy is emitted close to the end, leading to an explosion. Black holes of this mass should be exploding right now. Events such as this are being looked for, but so far none have been seen. There is also a problem called the 'black hole information paradox' associated with evaporation of a black hole. As mentioned above, almost all the information associated with matter falling into a black hole disappears. This was not a problem at first when it was thought that the lifetime of a black hole is infinite. The 'lost' information resides inside the black hole. However, the Hawking process does not take any information from the inside of the black hole, so where does this information go when the black hole evaporates? Loss of information turns out to be a serious problem for physics, and the community is currently divided on whether the information is actually lost.

6.4 The singularity at the center of a black hole

All the matter falling through the event horizon of a black hole should form a gravitational singularity at its center. Classically, this would be a one-dimensional point object where all the mass is concentrated into an infinitely small space and spacetime has infinite curvature. All physical laws break down in such an extreme environment. Quantum mechanics tells us that this could never happen since the size of the singularity will be the Planck length. Nevertheless, it takes a quantum theory of gravity (currently unavailable) in order to understand the nature of the singularity. Even so, various speculations about black holes and singularities have been put forward. As with many of the speculations surrounding string theory, these are on the boundary between physics and metaphysics. Among them is the 'white hole', a hypothetical region of spacetime from which matter and energy emerge from a singularity. In this sense, it is the reverse of a black hole, and possibly the place where the information lost in a black hole reappears. Such a solution of the General Theory of Relativity has been found, but there are no

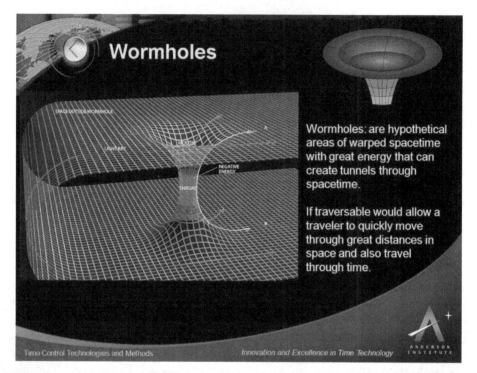

Figure 6.3. A wormhole connecting two regions of spacetime. Image courtesy of The Anderson Institute.

known physical processes by which a white hole could form. More intriguing is the 'Einstein–Rosen bridge' or 'wormhole', which is like a tunnel with its two ends at separate points in spacetime (figure 6.3). If such a thing existed, it could form a shortcut from one point in space to another very far distant point, allowing for travel at speeds much faster than light. As an alternative, it could also make time travel possible.

Chapter 7

Gravitational waves (the early years)

As mentioned in the section on general relativity, Einstein's equation predicted the existence of gravity waves just as Maxwell's equations 'predicted' the existence of electromagnetic waves, a.k.a. light. One of the major differences between these two forms of radiation is the extreme weakness of the force of gravity, which accounts for why it took 100 years after the publication of the theory for gravity waves to be directly detected. In fact, discouraged by the very small magnitude of the expected signal, nobody even attempted detection for the first 50 of those years. Then, in the period from 1967–69, Dr Joseph Weber of the University of Maryland published a series of three papers in *Physical Review Letters* (the physics journal of record for important new discoveries) making increasingly stronger claims for gravity-wave detection. The apparatus that he constructed, the 'Weber bars', consisted of cylinders of aluminum that were typically 30 inches in diameter and 30–60 inches long (figure 7.1). The bigger of these weighed 1350 pounds.

Weber's idea was that a passing gravity wave would alternately compress and expand the bar and the resulting signal could be amplified by the 'resonance phenomenon'. An analogy here is to the way the body of a violin or acoustic guitar amplifies the sound from a vibrating string. The amplification can be very large, but only at a specific 'resonant' frequency (or set of frequencies). Weber chose to look for gravity waves at a frequency of 1660 Hz, which he suggested might occur during a supernova collapse, and he 'tuned' his bars to resonate at that frequency. This is well within the range of human hearing, corresponding closely to the note emitted by the 72nd key (black) on a piano. Even with resonant amplification the signal was expected to be very small, so Weber bonded multiple 'piezoelectric sensors' to his bars. These devices convert compression and expansion to electrical signals. Of course, the bars had to be extremely well isolated from other sources of vibration, such as traffic on nearby roads for example. Weber also had to contend with signals resulting from the fact that the atoms in materials such as Weber bars are not at rest but vibrate because they are at a finite temperature. This 'thermal noise' was the

Figure 7.1. Joseph Weber and one of his bars. Dr Joseph Weber with gravity wave detector; copyright University of Maryland; University AlbUM; Special Collections and University Archives, University of Maryland libraries.

ultimate limiting factor to the sensitivity of Weber's apparatus. It was claimed that a change in the separation of the two faces of the bar by an amount approximately equal to a tenth of the radius of an atomic nucleus could be detected. In his first paper, Weber reported the detection of 10 events over a period of 2 years that were more than five times the thermal vibrations and therefore candidates for gravity waves, but other possibilities could not absolutely be ruled out. His second paper reported on the simultaneous detection of four events in two Weber bars separated by a distance of 2 km over a period of three months, and in the third paper he reported such 'coincidences' in bars separated by 1000 km, which seemed to rule out any explanation but gravity waves. This report generated a tremendous amount of excitement in the physics community and soon multiple scientists were building Weber bars, some with major improvements to the sensitivity of his instruments. Theoretical physicists also got into the act, calculating possible sources of gravitational radiation. Unfortunately these calculations showed that the amount of energy carried by Weber's events corresponded not to the collapse of a single supernova, as he had suggested, but instead to the complete destruction of thousands of massive stars. Furthermore, none of the newer instruments detected any events at all. By 1974, it was clear that there were problems with Weber's experiment. His refusal to ever acknowledge this criticism led to a serious decline in his reputation, which lasted until his death in 2000. Nowadays, however, he is recognized as the father of gravitational radiation experimentation because his pioneering work, though flawed, led others to design far more sensitive instruments that eventually resulted in the successful direct detection of gravity waves.

7.1 Gravitational waves (indirect detection)

The first successful detection of gravitational radiation was actually an indirect measurement. In 1974, Russell Hulse and Joseph Taylor of the University of Massachusetts-Amherst discovered pulsed radio waves which they identified as coming from a neutron star; in other words, a pulsar. However, the time between pulses (the period of the pulsar) varied regularly and repetitively. At times, it was shorter and later it was longer than the average. They soon recognized that this behavior was due to the fact that the pulsar was in an orbit around another object. By analyzing the details of the orbit, Hulse and Taylor were able to show that the other object was also a neutron star—the first discovery of a binary neutron star. Binary star systems are actually quite common in the Universe, but in this case both of the original massive stars collapsed and formed neutron stars. This was interesting enough, but the biggest surprise was yet to come. After longer-term observation of the system, it was found that the average period itself was slowly decaying due to the emission of gravitational radiation, at a rate that was exactly equal to expectations from Einstein's theory. The calculated intensity of the gravity waves, though very large, was far too small to be detected on Earth with the instruments available at the time. This indirect confirmation of the existence of gravitational radiation won Hulse and Taylor the 1993 Nobel Prize in Physics.

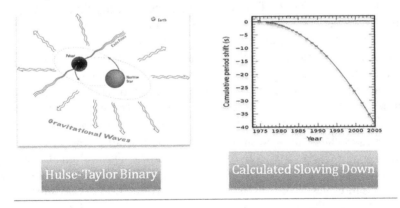

Figure 7.2. (Left) Credit: Shane L Larson. (Right) Public domain.

Chapter 8

The LIGO project

The progress in the theory of gravitational waves generated by Weber's claim of detection, as well as the indirect detection of gravity waves by Hulse and Taylor, caused a surge of interest in the 1970s in the possibility of their direct detection by instruments on Earth. It was soon realized that Weber bars had fundamental limits that couldn't be overcome and that a new type of detector, with a sensitivity perhaps 100 000 times greater, was needed. In the late 1960s, a concept for such a detector based upon a Michelson interferometer (chapter 2) showed considerable promise. The idea here was that a gravity wave passing through the device would alternately stretch and compress an arm of the interferometer. Also, because of the nature of the wave, the other arm would alternately be compressed and then stretched so that the maximum difference in the lengths of the two arms would be twice as big as the difference in either arm by itself. When light waves traveling through the two arms are brought together at a detector, the corresponding interference pattern would reflect the different path lengths seen by the beams (figure 8.1).

One important advantage of such a device is that it doesn't depend on resonant amplification of the signal so it is sensitive to a wide range of frequencies of gravitational radiation. However, to achieve the required sensitivity, the arms of the interferometer need to be at least 2–4 km long and the light beams must travel through a vacuum. Furthermore, the elements of the interferometer must be isolated as much as possible from vibrations, due to, for example, vehicles traveling on nearby highways or seismic waves from earthquakes. The technical challenges of building such a device are clearly extreme, and many experimental research groups worked throughout the 1970s on ideas to improve the sensitivity of interferometers. One very important theoretical development was the work of Kip Thorne and his group at Caltech and Vladimir Braginsky and his group at Moscow State University. In 1968, they showed that the heavy mirrors of a gravity-wave interferometer must actually be treated as if they formed a quantum system. Quantum mechanics is ordinarily thought of as applying only to the behavior of

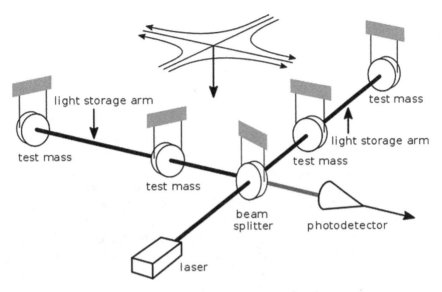

Figure 8.1. Design concept for a gravity-wave interferometer.

very small objects such as atoms. However, Braginsky pointed out that the effort to detect gravity waves required the measurement of extremely small displacements of the mirrors, much less than the size of an atom, so that the rules of quantum mechanics must apply. This was potentially devastating since the 'uncertainty principle' meant that the precise measurement of such a small displacement of a mirror would result in a disturbance to its momentum that would swamp the expected signal. Fortunately, Thorne and Braginsky were also able to show that there was a way around this quantum limitation via the concept of 'quantum non-demolition', that is, by the use of sensors that were designed so that the uncontrollable change in momentum of the mirror did not demolish information on its position. Kip Thorne gave an example of one such measurement in his 1994 book 'Black Holes & Time Warps'. This is the so-called 'stroboscopic quantum non-demolition measurement' (not the one actually used in LIGO). The idea is that, while the precise measurement of position at a given time gave an unknown 'kick' to the mirror, the wave is periodic and the position measured at exactly one (or two or three or more) periods after the initial kick would always be the same no matter the magnitude or direction of the change in momentum. Thorne and Braginsky worked on the actual implementation of the quantum non-demolition concept to gravity-wave detectors from 1968 until the 2000s.

It was recognized early on that constructing a giant interferometer of the required scale would be an expensive proposition, on the order of the 'big science' projects in high-energy physics. Ordinarily, such projects in the United States are funded by the Department of Energy (DOE). However, the LIGO (Laser Interferometer Gravity-Wave Observatory) was instead funded by the National Science Foundation (NSF) and ultimately became its most expensive project. Meticulous preparation was required to ensure the eventual success of the device. In 1980, the NSF funded the construction of prototypes at Caltech (40 m long) and MIT (1.5 m long) to test design concepts. The success of this exploratory work led to the formation of a joint Caltech/MIT planning group which further developed the concept, and in 1991 Congress approved the first year of funding for the construction of LIGO. Rather than just a single interferometer, it was decided to build two, one at Hanford, Washington and the other at Livingston, Louisiana, separated by about 3800 km. That way, it was possible to look for coincidences in the detection of signals from the two devices, thus reducing the probability of false events arising from local vibrations unrelated to gravity waves. In addition, the two sites were far enough apart that the time difference in detection of the two signals could be used to extract information on the direction of the incident wave. Civil construction of the two laboratories began in 1994 and was finished by 1998. The initial two interferometers were installed and commissioned by 2002, and the first (unsuccessful) searches for gravity waves were completed by 2010. This lack of success was not entirely unexpected. The original design concept envisioned the construction of interferometers based on proven technologies, which would then serve as test beds for the development of more advanced designs with higher sensitivity. It was understood at the time, based on theoretical calculations, that the initial LIGO would at best be sensitive to only the very most violent astrophysical events that could be imagined, such as a nearby merger of two black holes, and that a further increase in sensitivity by a factor of 10 or more would be necessary to ensure actual detection. Nevertheless, the initial experiments did provide a useful upper limit on the intensity of gravitational radiation.

Before discussing the developments that made the successful detection of gravity waves possible, it is helpful to talk about some of the properties of the original LIGO interferometers. Referring to the figure above, note the 'light storage' arms which are 4 km long. It turns out that this is far too small a distance to deliver the required sensitivity if the light waves from the laser only traverse the arms a single time. What isn't shown are 'Fabry–Perot' cavities in each arm near the beam splitter mirror. These devices act to reflect parts of each laser beam so they bounce back and forth 280 times before they are finally brought together at the photodetector, in effect making the interferometer 1120 km long! Note also the blue bars that indicate the suspension platforms for the mirrors, which hang from them via steel wires. These provide the required isolation from local vibrations. The mirrors themselves weighed 11 kg. Since light beams carry momentum, they cause the mirrors to recoil when reflected from them. This results in unwanted changes in the distance between the two mirrors. The heavier the mirrors, the smaller this effect. Another effect of the laser beams is that they heat up the mirrors, which can cause unwanted local deformations in its surface that can add to the 'noise' of the system. Heavy mirrors also reduce this effect. Then there's the laser itself. The original Michelson–Morley experiment didn't use lasers since they hadn't been invented at the time. The advantages of a laser light source are many. The light output is monochromatic to a high degree, which is very important in ensuring a clean interference pattern since light of different colors produces different interference patterns which may overlap. Laser light is 'coherent' with all the waves in step with each other, again aiding the clarity of the pattern. It's also highly directional so that it maintains its direction over the very large distances employed in the LIGO interferometers. For example, lasers have been used since the 1970s to accurately measure the distance to the Moon. After traveling an average distance of about 240 000 miles, the laser beam is only about 4 miles wide, and some of this is due to distortions caused by the Earth's atmosphere. Finally, lasers can produce very bright light. The LIGO lasers produce a light output power of 200 kW, at the limits of current technology. In comparison, the light output of a typical laser pointer is 5 mW, or 40 000 times weaker. However, in order to achieve the required sensitivity, even a much greater light output than that was required. Again, a technological trick like the Fabry–Perot cavities was necessary. So-called 'power recycling' mirrors were placed between the laser and the beam-splitter. The light passes through the power-recycling mirror and is split to travel into the two arms. After reflecting from the mirrors at the end of the arms, most of the light travels back to the recycling mirror where it is reflected back into the interferometer. The net effect is to increase the effective power of the beam by a factor of almost 4000 (see 'LIGO's interferometer') (figure 8.2).

Finally, a very important component of LIGO technology is the vacuum, which is maintained throughout the entire device including the 4 km long arms. Air in the system would have a number of deleterious effects. Like glass, air has an 'index of refraction' and therefore acts like a lens, changing the apparent distance between the mirrors. Even just a few molecules would have this effect and, since the air is at a finite (and possibly changing) temperature, its molecules are in constant motion causing fluctuations in the apparent path of the laser light that add 'noise' to the very weak signal expected from a gravity wave. In addition, air molecules would absorb

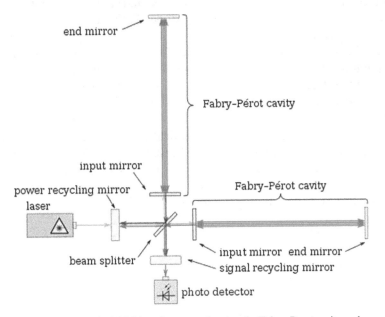

Figure 8.2. Partial schematic of the LIGO interferometer showing the Fabry–Perot cavity and power recycling mirror. In actuality, multiple such cavities and mirrors are used.

and scatter the laser light, and cause the mirrors to recoil unpredictably due to random collisions with their surfaces. The vacuum inside the LIGO interferometers is about one-trillionth of atmospheric pressure at sea level, surpassed for such a large device only by the vacuum in the beam pipes of the Large Hadron Collider.

As mentioned above, it was understood from the beginning of the LIGO laboratories that an increase in the sensitivity of the interferometers by at least a factor of ten was required to ensure the detection of gravity waves. As a result, construction of an 'advanced' LIGO was begun in 2008 while observations with the original instrument were still ongoing. Then, in 2010, the interferometers were shut down and installation of the improvements was begun. Major parts of that project had to do with the mirrors. The new mirrors have a mass of 40 kg, compared with 11 kg in the original design. This had a major effect on the recoil and thermal effects discussed above. In addition, a more complicated suspension (the 'quadruple pendulum') was installed to improve the vibration isolation. The initial design had only a single pendulum. Each stage of the quadruple pendulum adds further mechanical isolation from external vibrations. At the same time, the stainless steel suspension wires were replaced by non-thermally-conducting silica fibers, which have better thermal properties and thereby reduce the thermal noise of the system. Another major change was the introduction of 'active' vibration isolation of the mirrors. Sensors throughout the device detect vibrations and the information is fed back to actuators which counteract them, similar to the way in which noise-cancelling headphones work. This is part of the 'Feedback and Control' systems that continually monitor and control many aspects of the total LIGO apparatus. All these changes were finally completed in 2015 (but projects to further improve the performance of LIGO are still underway). Remarkably, the present device is sensitive to gravity waves having an amplitude that's 10 000 times smaller than the radius of a proton!

As befits any 'big science' project, LIGO science is carried out by a large collaboration of scientists, called the 'LIGO Scientific Collaboration' (LSC), which consists of over 900 scientists in 18 different countries. Furthermore, it's not the only gravity-wave detector on the planet. There is a similar facility in Italy (VIRGO) and a smaller version in Germany (GEO600). In addition, a Japanese facility (KAGRA) that uses advanced concepts such as cryogenically cooled optics (to reduce thermal noise) is currently under construction, and IndIGO (LIGO-India) is in the planning stage. Having several such detectors separated by very large distances is crucial for deducing the direction of a gravity wave, which in essence allows for the localization of the astrophysical source that generated it.

Chapter 9

Gravity wave astronomy

The first successful direct observation of gravitational radiation occurred on September 14, 2015, about 4 months after 'Advanced LIGO' achieved its initial design sensitivity goal and only two weeks after the beginning of the first official search campaign. It is designated as GW150914, based on 'gravity wave' and the date of observation (figure 9.1).

The same event (shown in the figure 9.1) was recorded at the Livingston site and 7 ms later at the Hanford site, consistent with the time for the wave to travel between the two locations at the speed of light. (The VIRGO detector in Italy was offline at the time and the GEO detector in Germany wasn't sensitive enough to observe the event.) The 'strain' in this figure represents the change in the distance between the mirrors of LIGO divided by the 4 km length of the arms, so the inferred displacement is roughly 1000 times smaller than the size of a proton. It can also be seen that the frequency of oscillation changed during the course of the event, from about 35 Hz at its beginning to 250 Hz at the end, within the range of human hearing (figure 9.2). This has been likened to the 'chirp' of a bird song, and in fact signals like this are referred to by physicists as 'chirps'.

The explanation for this behavior is that the gravity wave originates from two massive objects orbiting each other. The orbit for such a system gets smaller and smaller as energy is lost due to gravity-wave emission, and the orbital frequency subsequently increases until the objects fuse together, at which point gravity-wave emission ceases (except for the 'ringdown' of the resulting larger black hole) (figure 9.3). Animations of such a merger are shown here.

The details of the signal contain a wealth of information regarding the nature of the orbiting bodies. For example, see the 'reconstruction' image below. The lower panel in this image shows a reconstruction of the Hanford event using three different methods, all of which assume that the system consisted of a 30 solar mass black hole orbiting and eventually merging with one of 35 solar masses. The agreement with observation is excellent. The frequency of the gravity wave when it

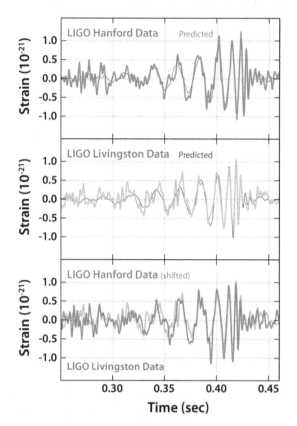

Figure 9.1. Signals for GW150914 (Hanford data shifted by 7 ms).

reaches its maximum amplitude (about 150 Hz) is twice the 75 Hz orbital frequency of the objects at that time, from which one can deduce that their centers were only about 350 km apart at the time of merger. Such small massive objects could only be black holes, and their calculated relative orbital speed increased from 20% to 60% of the speed of light during the course of the detected event. The reconstructions of the event showed that the total energy radiated in gravitational radiation was equivalent to 3 solar masses and the power emitted in the last 20 ms of the merger was 50 times greater than that of the light radiated during the same interval by all the stars in the visible Universe! Clearly this was a very violent astrophysical event, which again shows the great difficulty in detecting gravity waves and explains why they weren't directly detected until 100 years after Einstein's prediction. The gravitational wave energy emitted during a core-collapse supernova in another galaxy, thought by Weber to be the source of the signals he was seeing, would be very small compared with this event and even well beyond the capability of the present version of LIGO to detect. However, it would very likely be possible to detect gravity waves from a supernova event in the Milky Way. The last such event occurred in 1604, but the average interval between supernovae in other galaxies is about 300 years.

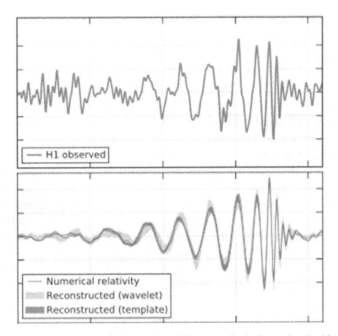

Figure 9.2. Reconstruction of GW150914 using three different methods. Reproduced with permission from Abbott B P *et al* 2016 *Physical Review Letters* **116** 061102. American Physical Society.

Figure 9.3. Orbiting massive objects (neutron stars or black holes) produce gravity waves. Credit: NASA/Tod Strohmayer (GSFC)/Dana Berry (Chandra X-Ray Observatory).

Following the breakthrough first direct observation of gravity waves, LIGO has detected six more events (plus one possible event) in two campaigns during 2015 and 2017. All but the last of these events (GW170817) involved black-hole mergers, based on the parameters of the gravity-wave signal. In this case, the orbiting objects were only 30 km apart at merger, and the frequency at maximum amplitude was approximately 1000 Hz compared with about 150 Hz for the black-hole mergers. GW170817 was also notable for a number of other reasons. First, it involved the merger of two neutron stars (1.3 and 1.5 solar masses) to form a black hole. The energy radiated in this event was far smaller than the others (0.025 solar masses vs 3 solar masses for GW150914), but it also occurred far closer to the Earth which is why it could be detected. Secondly, the VIRGO detector also saw this event so its location on the sky could be accurately determined from three measurements (two from LIGO and one from VIRGO). Finally, and most importantly, the precise location of the event allowed for its observation first by gamma-ray detection and later in optical telescopes. This was possible because the 'chirp' from the event was about 100 s long compared with fractions of a second for black-hole mergers, which gave time for 'multimode' detection involving both gravitational and electro-magnetic radiation.

Gamma rays were the first to be detected. They were of particular importance because of the connection to 'Gamma-ray bursts'. These were first seen in the late 1960s by satellites launched to detect nuclear explosions, but the observations remained classified until the early 1970s. It was subsequently shown that they come from outside the Milky Way because they are uniformly distributed across the sky. There are two classes of such events, distinguished by their time scales. The longer bursts, which can last for up to hours, are thought to result from the collapse of massive stars in supernova events. Other, shorter-term, events may last for only tens of milliseconds and these were thought to result from the merger of neutron stars. GW170817 showed this to be the case. Interesting questions are posed by the fact that the gamma-ray event began 1.7 s after the merger of the neutron stars. Several different explanations have been proposed for this delay, mostly involving the fact that electromagnetic radiation cannot initially penetrate the dense cloud of material ejected during the merger. However, this very short delay compared with the travel time of the gravitational and electromagnetic radiation (about 130 million years) places an upper limit of about 1 μm s^{-1} (10^{-6} m s^{-1}) on the difference between the speed of light and that of gravity, which is about 14 orders of magnitude better than any previous limit and eliminates some models of non-standard general relativity.

The detection of x-rays and ultraviolet, visible, and infrared light continued for days and weeks after the neutron-star merger, due to radioactive decay of the chemical elements produced during the event. The initial radiated power due to this effect is estimated to have been about 10 million times that of the Sun. An event of this type, dubbed a 'kilonova', was presumed to be due to the merger of two neutron stars but that had never before been directly observed. This was a very important confirmation of theories of the formation of the chemical elements. It is known that only the very lightest elements, up to Li, were produced in the Big Bang at the formation of the Universe. Elements up to the Fe region are produced by fusion of

Figure 9.4. The Crab Nebula: a supernova remnant. Image: NASA.

lighter elements within the cores of massive stars, and then distributed into the cosmic environment when these undergo supernova events. An example is the Crab Nebula, the remnant of a supernova observed by Chinese astronomers in the year 1054 (figure 9.4). Elements produced within the precursor star have been observed in the remnant. As mentioned previously, there is a neutron star hidden within this nebula, as well as regions of new-star formation.

Stars of lesser mass never reach this stage and the elements they produce remain locked up within white dwarves, unless they are in binary systems. In that case, hydrogen gas from the companion star may fall onto the surface of the white dwarf, which could then eventually achieve the conditions for a thermonuclear explosion. This is a so-called 'nova' event, which can typically produce and distribute elements up to the Mg region. The difficulty comes in understanding the production of elements heavier than Fe. Fusion of these heavier elements is 'endothermic', i.e. it requires energy input in order to proceed (as opposed to the 'exothermic' fusion of lighter elements, which produces energy). Therefore, synthesis of elements beyond Fe must occur in highly-energetic astrophysical events, such as in the midst of a supernova explosion. However, for the heaviest elements even this is not enough. Theories of the formation of such elements, which have many more neutrons than protons in their nuclei, require high-energy environments that also have a large excess of available neutrons. These are not easy to find. Neither the cores of massive stars nor supernova 'fireballs' feature large numbers of neutrons, so the theories of the production of heavy elements had no natural astrophysical sites until the concept of the 'kilonova' was introduced. The region around merging neutron stars is clearly a high-energy, neutron rich environment, an ideal location for the production of the heaviest elements. GW170817 provided the first direct evidence of a kilonova in action. The production of heavy elements totaling approximately 5% of the mass of the Sun was observed. Though this might seem like only a small amount of matter, it can account for all of the heavy elements in the Galaxy if the number of these kilonova events is as expected. Future gravity-wave observations will be required to see if this is the case.

Analysis of the data obtained from GW170817 continues to give remarkable new insights into several different areas of physics. For example, a recent manuscript (November, 2018) showed that the properties of its gravity-wave signal provided the first real test of Einstein's general relativity theory in the immensely strong gravitational field near orbiting neutron stars. The theory passed with flying colors. It was also possible to place tight constraints on 'leakage' of the effects of gravity into large extra dimensions, as proposed in certain theories of 'quantum gravity' which attempt to account for the extreme weakness of gravity compared with other forces in the Universe.

LIGO is currently down for maintenance and upgrading in 2018 but a third campaign is expected to begin in the Spring of 2019. We are at the threshold of a new kind of astronomy in which information obtained from electromagnetic and gravitational radiations can be combined to form a much better picture of objects such as neutron stars, and of the production of the chemical elements. Already, optical studies of the aftermath of the GW170817 event have been shown to be consistent with the hypothesis that the heaviest elements are produced during neutron-star mergers. In fact, the amount of information that was obtained from just this single event is truly astounding, and astrophysicists are eagerly looking forward to the treasure trove of data that will be obtained from future observations of similar events. On the other hand, black holes are by their very nature rather difficult to see, only revealing themselves in cases where infalling matter is heated to the point that it emits radiation in various parts of the electromagnetic spectrum. Gravitational radiation from black-hole mergers will give a new picture of this part of the Universe that has heretofore essentially been invisible. Clearly, the future of gravitational-wave astronomy is extremely bright, and it will be at the forefront of observations of the Universe for a very long time.

9 781643 274232